The publishing house tradition has created the series **TREDITION CLASSICS**. It contains classical literature works from over two thousand years. Most of these titles have been out of print and off the bookstore shelves for decades.

The book series is intended to preserve the cultural legacy and to promote the timeless works of classical literature. As a reader of a **TREDITION CLASSICS** book, the reader supports the mission to save many of the amazing works of world literature from oblivion.

The symbol of **TREDITION CLASSICS** is Johannes Gutenberg (1400 – 1468), the inventor of movable type printing.

With the series, tradition intends to make thousands of international literature classics available in printed format again – worldwide.

All books are available at book retailers worldwide in paperback and in hardcover. For more information please visit: www.tredition.com

tradition was established in 2006 by Sandra Latusseck and Soenke Schulz. Based in Hamburg, Germany, tradition offers publishing solutions to authors and publishing houses, combined with worldwide distribution of printed and digital book content. tradition is uniquely positioned to enable authors and publishing houses to create books on their own terms and without conventional manufacturing risks.

For more information please visit: www.tredition.com

The Art of Making Whiskey So As to Obtain a Better, Purer, Cheaper and Greater Quantity of Spirit, From a Given Quantity of Grain

Anthony Boucherie

Imprint

This book is part of the TREDITION CLASSICS series.

Author: Anthony Boucherie
Cover design: toepferschumann, Berlin (Germany)

Publisher: tredition GmbH, Hamburg (Germany)
ISBN: 978-3-8491-8417-9

www.tredition.com
www.tredition.de

Copyright:
The content of this book is sourced from the public domain.

The intention of the TREDITION CLASSICS series is to make world literature in the public domain available in printed format. Literary enthusiasts and organizations worldwide have scanned and digitally edited the original texts. tredition has subsequently formatted and redesigned the content into a modern reading layout. Therefore, we cannot guarantee the exact reproduction of the original format of a particular historic edition. Please also note that no modifications have been made to the spelling, therefore it may differ from the orthography used today.

CONTENTS

- PREFACE.
- CHAPTER I.
- CHAPTER II.
- CHAPTER III.
- CHAPTER IV.
- CHAPTER V.
- CHAPTER VI
- CHAPTER VII.
- CHAPTER VIII.
- CHAPTER IX.
- CHAPTER X.
- CHAPTER XI.
- CHAPTER XII.
- CHAPTER XIII.
- CHAPTER XIV.
- CHAPTER XV.
- THE ART OF MAKING GIN

UNITED STATES OF AMERICA,

District of Kentucky, to wit:

Be it remembered, That on the 10th day of December, in the year of our Lord, 1818, and the forty-third year of the Independence of the United States of America, came Anthony Boucherie, of the said district, and deposited in this office, a copy of the title of a book, the right whereof he claims as author and proprietor, in the words and figures following, viz:

"The Art of making Whiskey, so as to obtain a better, purer, cheaper and greater quantity of Spirit from a given quantity of Grain: Also, the art of converting it into Gin, after the process of the Holland Distillers, without any augmentation in the price. – By Anthony Boucherie:"

In conformity to the act of Congress of the United States, entitled "An act for the encouragement of learning, by securing the copies of maps, charts and books to the authors and proprietors of such copies during the times therein mentioned." And also to an act, entitled "An act supplementary to an act, entitled an act for the encouragement of learning, by securing the copies of maps, charts and books to the authors and proprietors of such copies, during the times therein mentioned, and extending the benefits thereof to the arts of designing and etching historical and other prints."

JOHN H. HANNA,

Clerk of the District of Kentucky.

[Pg 3]

TO THE

HONOURABLE LEGISLATURE

OF THE

STATE OF KENTUCKY.

Gentlemen of the Senate,
and of the House of Representatives,

An immense and most fertile country, a republic where every individual enjoys the most unbounded freedom; such are the advantages which characterise the United States of America, and render them the asylum of the oppressed Europeans. I was one of the number, and as early as January, 1808, congress enacted a law dispensing me with the usual term of two years residence, for obtaining a patent.

It is the duty of every citizen to contribute to the progress of useful knowledge, for the benefit and prosperity of his native or adopted country. It is under that point of view that I now publish The Art of Making Whiskey, *so as to obtain a greater quantity of Spirit from a given quantity of Grain; the spirit thus obtained being purer and cheaper. Also, the Art of converting it into Gin, according to the process of the Holland Distillers, without making it dearer.*

This next paragraph is incomplete

Give me leave, gentlemen, to publish this little work under the patronage of the enlightened Legislature of the state which I have chosen for my residence is undoubtedly of a general utility fo – [Pg 4] but more particularly an agricultural state, such as this, where every thing that contributes to the success of agriculture, adds to the welfare of the commonwealth. It is therefore to promote that desirable end, that I hereby renounce all the privileges granted me eight years ago, for the distiller's apparatus, of which I give here a description. I invite all distillers to use it the more confidently, as a long experience has proved to me its utility. In describing the art of converting Whiskey into Gin, according to the process of the Holland Distillers, I flatter myself, that I give a greater value to a national production usually neglected throughout the continent, and which will be the principle of a considerable produce. Henceforth the Gin of the United States will be an important article of exportation for their outward trade, as well as for home consumption.

Receive, gentlemen, the
Assurances of my
Profound Respect,
A. BOUCHERIE.

[Library stamp: IMPERFECT IN ORIGINAL]

[Pg 5]

PREFACE.

The most usual drink in the United States, is whiskey; other spirituous liquors, such as peach and apple brandy, are only secondary, and from their high price and their scarcity, they are not sufficient for the wants of an already immense and increasing population. As to wine, in spite of all the efforts and repeated trials made to propagate the grape-vine, there is as yet no hopes, that it may in time become the principal drink of the Americans.

To turn our enquiries towards the means of bringing the art of making whiskey to greater perfection, is therefore, to contribute to the welfare of the United States, and even to the health of the Americans, and to the prosperity of the distiller, as I will prove in the sequel.

The arts and sciences have made great progress; my aim is to diffuse new light on every thing that relates to the formation of spirituous liquors that may be obtained from grains. Most arts and trades are practised without principles, perhaps from the want of the means of information. For the advantage of the distillers of whiskey, I will collect and offer them the means of obtaining from a given quantity of grain, the greatest possible quantity of spirit, purer and cheaper than by the usual methods. I shall then proceed to indicate the methods of converting whiskey into gin, according to the process of the Holland Distillers, without heightening its price.

If the principles hereafter developed are followed, the trade of distiller will acquire great advantages, that will spread their influence on agriculture, and consequently on commerce in general.

[Pg 7]

THE ART OF MAKING WHISKEY, &c.

CHAPTER I.

OF SPIRITUOUS LIQUORS, OR SPIRITS.

Spirituous liquors are the produce of vinous ones, obtained by the distillation of these last. The art of making wine is of the remotest antiquity, since it is attributed to Noah; but that of distilling it, so as to extract its most spirituous part, dates only from the year 1300. Arnand de Villeneuve was the inventor of it, and the produce of his Still appeared so marvellous, that it was named Aqua-Vitæ, or *Water of Life*, and has ever since continued under that denomination in France; Voltaire and reason say that it might, with far more propriety, be called *Aqua-Mortis*, or Water of Death.

This liquor, called in English, *Brandy*, received from the learned the name of *Spirit of Wine*; time improved the art of making it still stronger by concentration, and in that state it is called *Alcohol*.

All spirit is the distilled result of a wine, either of grapes, other fruits, or grains; it is therefore necessary to have either wine, or any vinous liquor, in order to obtain spirits. [Pg 8]

CHAPTER II.

OF THE FORMATION OF VINOUS LIQUORS WITH GRAINS, IN ORDER TO MAKE SPIRITS.

The art of extracting wine from the juice of the grape, not being the object of this book, I shall confine myself to what is necessary and useful to the distillers of whiskey; it is therefore of the vinous liquor extracted from grains, that I am going to speak.

The formation of that kind of liquor is founded upon a faculty peculiar to grains, which the learned chymist, Fourcroy, has called *saccharine fermentation*. Sugar itself does not exist in gramineous substances; they only contain its elements, or first principles, which produce it. The saccharine fermentation converts those elements into sugar, or at least into a saccharine matter; and when this is developed, it yields the eminent principle of fermentation, without which there exists no wine, and consequently no spirit.

Grains yield two kinds of vinous liquors, of which the distiller makes spirit, and the brewer a sort of wine, called *beer*. From a comparison of the processes employed to obtain these two results, it will be found that the brewer's art has attained a higher degree of perfection than that of the distiller. They both have for their object to obtain a vinous liquor; but that of the brewer is, in reality, a sort of wine to which he gives, at pleasure, different degrees of strength; while that of the distiller is scarcely vinous, and cannot be made richer. I will give a succinct exposition of their two processes in order that they may be compared.

OF THE ART OF BREWING.

The art of brewing consists:

1st. In the sprouting of a proportion of grain, chiefly [Pg 9] barley. This operation converts into a saccharine matter, the elements of that same substance already existing in grains.

2dly. In preparing the *wort*. For that operation, the grain, having been previously ground, is put into a vat, which is half filled up with water; the rest is filled up at three different times with hot water—the first at 100°, the second at 150°, and the third at 212°, which is boiling water. The mixture is strongly stirred each time that it is immersed. By this infusion, the water lays hold of the sweet principles contained in the grain.

3dly. The wort thus prepared, the liquor is filtrated, in order to separate it from the grain, and then boiled until reduced to one half, in order to concentrate it to the degree of strength desired. In that state, 40 gallons of wort contain the saccharine principles of 200 wt. of grain.

4thly. The wort, thus concentrated, is drawn off in barrels, which are kept in a temperature of 80° or 85°. The yeast is thrown into it to establish the fermentation, and in a short time beer is made, more or less strong, according to the degree of concentration, and more or less bitter, according to the greater or lesser proportion of hops put into it.

Such are, in a concise view, the proceedings of the brewer. Let us proceed to those of the distiller of whiskey.

OF THE DISTILLER OF WHISKEY.

Whiskey is made either with rye, barley, or Indian corn. One, or all those kinds of grains is used, as they are more or less abundant in the country. I do not know how far they are mixed in Kentucky; but [Pg 10] Indian corn is here in general the basis of whiskey, and more often employed alone.

I have ascertained, in the different distilleries which I have visited in the United States—

1stly. That, in general, the grain is not sprouted. I have, however, seen some distillers who put 10lbs. of malt into a hogshead of fermentation containing 100 gallons, which reduces it to almost nothing.

2dly. That they put two bushels of ground grain into a hogshead of fermentation containing 100 gallons, filled up with water.

3dly. They had a ferment to determine the fermentation, which, when finished, yields two gallons of whiskey per bushel of grain, and sometimes ten quarts, but very seldom. I do not know whether those results are exact; but, supposing them to be so, they must be subject to great variations, according to the quality of the grain, the season, the degree of heat, of the atmosphere, and the manner of conducting the fermentation. From my analysing the different sorts of grains, I know that Indian corn must yield the most spirit.

From the above proportions, it results, that 100 gallons of the vinous liquor of distillers yield only 4 gallons of whiskey, and very seldom 5; that is, from a 25th to a 20th. It is easy to conceive how weak a mixture, 25 parts of water to one of whiskey, must be; thus the produce of the first distillation is only at 11° or 12° by the areometer, the water being at 10°. It is only by several subsequent distillations, that the necessary concentration is obtained, to make saleable whiskey. These repeated operations are attended with an increased expense of fuel, labor, and time.

Such are the usual methods of the whiskey distillers. Before we compare them with those of the brewer, [Pg 11] let us examine the nature of fermentation, and what are the elements the most proper to form a good vinous liquor: thence we shall judge with certainty, of those two ways of operating.

CHAPTER III.

OF FERMENTATION.

"Fermentation is a spontaneous and intestine motion, which takes place amongst the principles of organic substance deprived of life, the maximum of which always tends to change the nature of bodies, and gives rise to the formation of new productions."

Bouillon la Grange. – Manual of a Course of Chymistry.

Fermentation has long since been divided into *spirituous*, *acid*, and *putrid*.

It is only since the revival or new epoch of chymistry, that the learned have been occupied in researches on fermentation. I was the first who gave a new hint on this important part of natural philosophy, in 1785. It was then held as certain, that the saccharine substance was the principle of spirituous fermentation. A series of experiments enabled me to demonstrate the contrary, for I obtained a well crystallized sugar by the fermentation of a substance which produces none by any other means.

In September, 1785, I read a memoir to the Academy of Sciences, at Paris. In that memoir I developed my theory. That learned body nominated four commissioners, for the purpose of examining my operations, and sanctioned my discovery by a report, in [Pg 12] which it was acknowledged that I had discovered a new truth, and ordered the insertion of my memoir in the collection of those of the Foreign Associates. I attributed the principle of the spirituous fermentation to the mucilaginous substance. This has been since demonstrated, by attentively observing that it always begins with a motion of acid fermentation, which is produced by the mucilaginous substance. The European chymists have since reasoned upon fermentation; each of them has produced a new system; none have been able to bring it to a regular demonstration; and the learned Gay Lussac has said, that fermentation is one of the most mysterious operations of chymistry. Be that as it may, there are facts that are ascertained: let us endeavor to investigate them, that we may derive from them all the information which is necessary to us.

It is incontestable that spirits are produced by the saccharine substance. Grains, however, supply it, although they are not sensibly sweet. This has made me suspect that the fermentation is at first saccharine, which produces the sweet substance that is necessary for the formation of spirit. It is thus that, by a series of internal motions, the fermentation causes the formation of the spirit to be preceded by a slight production of acid; that it transforms the vinous liquor into vinegar, which the same fermentation changes in time into an animal substance, destroyed in its turn by the putrid fermentation. Such are the progressive changes operated by this all-disorganizing phenomenon, and the unerring march of nature to bring back all substances to their respective elements.

The necessary conditions for the formation of vinous fermentation, are —

[Pg 13]

- 1st. The presence of the saccharine substance.
- 2dly. That of a vegeto-animal substance, commonly called ferment, and soluble in water.
- 3dly. A certain quantity of water.
- 4thly. A temperature of 70° to 75°.
- 5thly. A sufficient mass.

When these are obtained, in a short time the liquor becomes turbid; it bubbles, from the disengaging of the carbonic acid gaz, and the heat increases considerably. After some days, these impetuous motions subside; the fermentation ceases by degrees; the liquor clears up; then it emits a vinous smell and taste. As soon as it ferments no more, it must be distilled. However, some distillers have asserted that a greater quantity of spirit is obtained when the liquor has acquired a certain degree of acidity. Others are of opinion that it must be distilled as soon as it is calm. I am of this opinion, because the acid can only be formed at the expense of a little of the spirit, which is one of the principles of the acetous acid. Besides, the longer the liquor remains in a mass, the more spirit is wasted by evaporation.

CHAPTER IV.

OF THE PROPORTIONS OF THE ELEMENTS NECESSARY TO FORM A GOOD VINOUS LIQUOR.

What are the proportions of the elements necessary to form a good vinous liquor?

We owe the important knowledge of those proportions to the celebrated and unfortunate Lavoisier, who [Pg 14] has proved, by the most accurate experiments, that there must be

100	parts of dry sweet substance, or sugar
400	parts of water
10	parts of ferment, or liquid yeast, which is reduced
— —	to 8 7-10ths of dry matter.

510 parts in the whole, which produce 57 parts of dry alcohol; that is, containing no more water than is necessary to its formation, and consequently as strong as it can be. Let us dwell for a moment upon the proportions just pointed out, and especially upon their result, which exceeds any thing that has ever been obtained. Supposing the weight of each of those parts to be one pound, we shall have

100	lbs.	of dry sweet Substance, or sugar
400	do.	of water
10	do.	of liquid ferment
— —		
510	pounds	in the whole.

100	lbs.	of sugar is the quantity required to make 12½ gallons of sirup, composed of 8lbs. of sugar and 8lbs. of water per gallon,	12½	galls.
400	lbs.	of water, at 8lbs. per gall. make	50	"

The produce will be 57lbs. of dry alcohol.

TR: Poor quality made it difficult to verify the above numbers and so noted with an asterisk

A vessel containing one ounce of water, filled up with this alcohol, weighs only 16dwts. and 16grs. From this report, it appears that the specific weight of the alcohol is, to the weight of the water, as 20 to 24; that is, that water weighs 1/5 more than alcohol. If the 57lbs. thus obtained were only water, it would only represent 7-1/8* gallons; but being alcohol, it weighs 1/6* less, and consequently gives 7-1/8 gallons more, the sixth of this quantity, (to wit:) 1-1/6* gallons, which, added to 7-1/8*, make 8-7/24 gallons.

[Pg 15]

But 1 gallon of dry alcohol, extended in 2 gallons of water, gives 3 gallons of liquor at 19°, which is called Holland, or first proof; a produce surpassing all what has been hitherto known to the distillers. I will prove it by an example: 1 gallon of molasses yields only 1 gallon of rum, at 19°, to the rum distiller; still, molasses is a true sirup, composed of 8lbs. of sugar, or sweet matter, more fermentable than sugar. 12½ gallons of molasses, representing 100lbs. of dry sweet matter yield consequently 12½ galls. of rum, Holland proof, which is only half the produce obtained by Lavoisier; an immense difference capable of exciting the emulation of all distillers, as it proves the imperfection of the art.

What are the causes of such a dissimilarity of product? We must seek for them.

1st. In the difference of the strength of the vinous liquor. Lavoisier employed only 4 parts of water to 1 part of dry sugar. The rum distiller usually puts 10 gallons of molasses to 90 gallons of water, or the residue of the preceding distillations.

10 galls. molasses contain

80 lbs. of sweet matter.

90 gallons of water weigh 720lbs.; therefore the proportion is, one part of sweet matter to 9 parts of water—whilst that indicated by Lavoisier is only 4 parts of water to 1 part of sugar. [a]

It is obvious how much richer this last must be, and that the fermentation thus produced has an energy far [Pg 16] superior to the other. Thence results a rapid production of spirit, operated in a short time; whilst that of the rum distiller languishes more or less, and a slow fermentation wastes part of the spirit which it produces, even as it is forming.

2dly. Bodies evaporate in proportion to the extent of their surface. One hogshead of 100 gallons, should contain, according to Lavoisier's composition, the elements of 50 gallons of spirit, at 19°; whilst that of the rum distiller contains only 12. Now, as every fermentable liquor requires open vessels, the hogshead of the rum distiller loses as much spirit as that of Lavoisier: hence it is plain how far the above proportion operates to the disadvantage of the fermer.

3dly. Another source of loss arises in the distilling vessels themselves. Nothing is more imperfect than the stills of a whiskey distillery. Lavoisier's were so perfect, that he made the analysis and the synthesis in the most delicate operations. [b] The vessels of the whiskey distillers, far from being hermetically closed, allow the spirit to evaporate through every joint. And this is not all: corroded by the acetous acid, they are full of small holes, particularly in the cap, where all the vapors collect themselves, as in a reservoir. It is easy to conceive with what rapidity they escape, which occasions a considerable waste of liquor. In proof of the truth of this observation, we may refer to the smell of whiskey, so strongly perceivable on the roads leading to a distillery, and preceeding from no other cause than that liquor wasting out of bad vessels, to the great loss of the distiller. [Pg 17]

4thly. A fourth cause of loss arises from the worm of the still. However careful in keeping the surrounding water cool, there is always one portion of vapor not condensed. This is made more sensible in the winter, when the cold of the atmosphere makes every vapor visible; upon examination, it will be seen that the running stream of liquor is surrounded with it. In my description of my apparatus, I give the means of obviating that evil.

To these several causes, may we not add another? May not the production of spirit be in a ratio to the richness of the fermenting liquor? It is certain, that in every spirituous fermentation there is a portion of the sweet matter which remains undecomposed and in its original state. Lavoisier found that it was 4.940; that is, nearly 5 parts in 100. It may possibly be the same in a weaker liquor; which would increase the loss, in the inverse ratio of the density of the liquor. Such are the causes to which I attribute the great superiority of Lavoisier's products; and from those observations I thought I could establish the fabrication of whiskey upon new principles.

CHAPTER V.

A COMPARISON OF THE PROCESSES OF THE BREWER WITH THOSE OF THE WHISKEY DISTILLER.

From the experiments of one of the most learned chymists of Europe, it has been demonstrated, that the proportions the most advantageous to the formation of a good vinous liquor, are, one part of dry sweet substance to four parts of water; that is, that the sugar [Pg 18] must form one fifth of the whole. We have, moreover, seen that 100lbs. of dry sweet matter gave 25 gallons of spirit 19°, which comes to 4lbs. of sugar per gallon.

We shall make use of that scale in comparing the processes of the brewer with those of the whiskey distiller.

Supposing the bushel of grain to weigh 50 pounds, and that it gives 2 gallons of whiskey at 19°, each of which gallons is the product of 4lbs. of sugar; then the strong beer which contains in 40 gallons the sweet matter of 200lbs. of grain, contains the elements of 8 gallons of spirit, or 32lbs. of dry sweet substance; and as the 40 gallons of this beer weigh 320lbs. the 32lbs. of sugar form only one-tenth of it, which is one half of Lavoisier's proportions.

Those of the distiller of whiskey are 100lbs. of grain to 100 gallons of water, or thereabouts: 100lbs. of grain contain only 16lbs. of dry sweet matter: therefore, as the 100 gallons of vinous liquor weigh 800lbs. the 16lbs. of sugar form only its fiftieth part.

Thence is seen how inferior the proportions of the whiskey distiller are to those of the brewer, and how far they are from good theory. But the brewer aims only at producing a sort of wine, and succeeds; while, the distiller wants to make spirit, and only obtains it in the manner the most expensive, and opposed to his own interest.

CHAPTER VI

DEFECTS IN THE USUAL METHOD OF MAKING WHISKEY.

1st. The most hurtful of all for the interests of the distillers, is undoubtedly the weakness of the vinous [Pg 19] liquor. We have seen that the proportion of spirit is in a ratio to the richness of the fermenting liquor; that Lavoisier, by putting one-fifth of the mass of dry sugar, obtained twice as much spirit as the rum distiller, who puts in the same quantity, but drowns it in water. From those principles, which are not contested, the distiller, whose vinous liquor contains only one-fiftieth part of sweet matter, obtains the less spirit, and loses as much of it as he gets.

2dly. Another defect is joined to this: bodies are dissolved by reason of their affinity with the dissolving principle; the mucilaginous substance is as soluble in water as the saccharine substance. A mass of 100 gallons of water having only 16lbs. of sugar to dissolve, exerts it's dissolving powers upon the mucilaginous part which abounds in grains, and dissolves a great quantity of it. There results from that mixture, a fermentation partaking of the spirit and the acid, and if the temperature of the atmosphere is moderate, the acid invades the spirit, which is one of its principles: nothing remains but vinegar, and the hopes of the distiller are deceived.

Some distillers have been induced, by the smallness of their products, to put in their stills, not only the fluid of the liquor, but the flour itself. Hence result two important defects. 1st. The solid matter precipitates itself to the bottom of the still, where it burns, and gives a very bad taste to the whiskey. In order to remedy this inconvenience, it has been imagined to stir the flour incessantly, by means of a chain dragged at the bottom of the still, and put in motion by an axis passing through the cap, and turned by a workman until the ebullition takes place. This axis, however well fitted to the [Pg 20] aperture, leaves an empty space, and gives an issue to the spirituous vapors, which escaping with rapidity, thereby occasion a considerable loss of spirit.

3dly. The presence of the grain in the still, converted into meal, is not otherwise indifferent. It contains a kind of essential oil, more or less disagreeable, according to its nature; which distils with the spirit. That of Indian corn, in particular, is more noxious than that of any other grain; and it is the presence of meal in the stills, which causes the liquors obtained from grains to be so much inferior to that of fruits.

4thly. There is a fourth defect, at which humanity shudders, and which the laws ought to repress. Vinous liquors are more or less accompanied with acetone acid, or vinegar; but those proceeding from grain contain still more of this acid. The stills are generally made of naked copper; the acid works upon that metal, and forms with it the *acetate of copper*, or verdigrise, part of which passes with the whiskey. There is no distiller, who, with a little attention, has not observed it. I have always discovered it in my numerous rectifications, and at the end of the operation, when nothing more comes from the still but what is called the sweet oil of wine. An incontestable proof of this truth is, that as the stills of the distillers are of a green color in their interior part; that they are corroded with the acid, and pierced with numberless little holes, which render them unfit for use in a very short time. It is easy to conceive how hurtful must be the presence of verdigrise to those who make use of whiskey as a constant drink: even those who use it soberly, swallow a slow poison, destructive of their stomach; while to those who abuse it, it produces a rapid death, which would [Pg 21] still be the consequence of abuse, if the liquor was pure, but is doubly accelerated by the poison contained in the whiskey. It is easy to remedy so terrible an evil. The acetous acid has no action upon tin. By tinning the stills, the purity of the liquor will be augmented, and the distilling vessels, already so expensive, will be longer preserved. This operation must be renewed every year. The worms must likewise be tinned, if they are copper; but they are better of tin, or of the purest pewter.

Such are the defects of the present method of distilling whiskey. Having exposed them, I must present the means of bringing to perfection the fabrication of a liquor of such general use.

CHAPTER VII.

DESCRIPTION OF THE PROCESS THE MOST ADVANTAGEOUS TO MAKE WHISKEY.

TR: The next 2 paragraphs were cut short, noted with [*]

As it is demonstrated that the spirit is the more abundant in proportion to the richness of the vinous liquor,* it is therefore necessary to enrich that of the distillery* which is so deficient in that respect. An exposition of* my processes will point out the means I employ to attain* that end. A large whiskey distillery should be* able to make 100 gallons per day, or three barrels* making altogether that quantity.

One gallon of spirit being the produce of 4 pounds* of dry saccharine matter, we must therefore have 400 pounds of this substance for the 100 gallons we wish to obtain.

If 1 bushel of grain gives 2 gallons of whiskey, there must be 50 to obtain a daily result of 100 gallons. I [Pg 22] take Indian corn as the basis of the fabrication, as that of all the grains which yields the most. For, from my method, whatever grain is employed, the spirit is equally pure.

I divide the still house into three different rooms, to wit:

One for Infusion;

One for Fermentation;

One for Distillation.

CHAPTER VIII.

THE ROOM OF INFUSION.

It is here that the liquor destined to make whiskey, should be prepared, and made rich enough to procure a good fermentation. To this effect, there must be a mill with a vertical stone, moved by a horse, or any other means of motion. Those mills are too well known for me to describe them more amply. The corn must be coarsely ground, so as scarcely to be broke into three or four pieces: consequently the stone must not be too heavy, for, at all events, the grain had better be too coarse than too fine. That mill should be placed in the infusion room, so as not to keep it dirty, nor to be too much in the way. It must grind, or rather break, 50 bushels per day.

There must be a square kettle, 4 feet broad, 5 feet long, 1 foot deep. The kettle must be made in sheets of copper, one line thick, at least: the bottom, although flat, should have a slight swell inside, so as to avoid the expansion of the metal outside, from the action of the fire. This kettle must be placed upon a brick fur [Pg 23] nace, so that the longest parts should bear forwards, and the other against the chimney, from which it must be separated by a brick wall eight or nine inches. The sides, around which there must be a space to walk freely, should be supported by a wall 1½ feet deep; the fore part upon such a wall, in the middle of which is an iron door, fifteen inches square, in an iron frame, through which the fuel is introduced.

The kettle is mounted upon the furnace, so as to bear upon the four walls about 4 inches, and rests upon a bed of clay, which must leave no passage to the action of the fire; it is lined externally with bricks, and must have a pipe on one of its sides, to draw off the liquor.

Under the kettle, 15 inches from the bottom, is a flue for the heat, running through all its length. It is 2½ feet wide at bottom, extending like a fan at the top, about 6 inches on each side, so that the flame may circulate in all the breadth of the kettle.

On the fore part of this flue, facing the door, is a hearth, occupying all its breadth, and 2 feet long. The rest of the flue is paved with bricks, and rises insensibly 4 inches towards the chimney, in which it opens by two holes, 1½ inches wide, 8 or 9 inches high.

Immediately under the hearth, is a mash hole 4 feet deep, occupying all its capacity, and projecting 2 feet forward. This opening is necessary to keep up a free circulation of air, and to take up the ashes. It should be covered with strong boards, not to hinder the service of the kettle. The hearth is made with an iron grate, more or less close, according to the nature of the fuel; if for wood, the bars must be about two inches apart; if for coals, half an inch is sufficient. The furnace must be built with care. The parts most exposed to the ac [Pg 24] tion of the fire must be built with soft bricks and potters' clay: soap stone would be preferable, if easy to procure. The brick separating the kettle and chimney, must be supported with flat bars of iron, as well as the part over the door.

CHAPTER IX.

USE OF THE KETTLE.

The kettle is destined to make the infusion of the grain, and boil it so as to convert it into wort. By that operation I make the liquor richer, which I intend for fermentation, and bring it to divers degrees of strength.

I put into the kettle 100 gallons of water, and 4 bushels of corn, broken, as I said before, at the mill. I light a small fire, which I increase gradually, until the water begins to boil; during that time, the grain is stirred with a paddle. As soon as the ebullition is established, the grain is taken up with a large skimmer, and put to drain into a large basket hanging over the kettle; and when the grain has been totally taken up, the fire is increased so as to bring the water to boil again, until reduced to two-fifths, which degree of concentration is not rigorous, and the distiller may augment it as his experience shall direct. When thus concentrated, the liquor is drawn off through the pipe, and received into a tub or vat containing 130 or 140 galls.

100 gallons more of water are put into the kettle, with 4 bushels of corn; the fire conducted slowly, as before, until the degree of ebullition; the corn is taken off, and the liquor concentrated in the same proportions; then drawn off as above, in the same tub. [Pg 25]

The same operation is repeated for the third time; the three united liquors are slightly stirred, and, still warm, transported into one of the hogsheads of fermentation, which it nearly fills up.

As there must be four of these hogsheads filled up daily, the work at the kettle must be kept going on, without interruption, until that quantity is obtained, which may be done in about twelve hours. The grain which has been drained is carried to dry, either in the open air, or in a granary, and spread thin. When dry, it is excellent food for cattle, and highly preferable to the acid and fermented mash, usually used by distillers to feed cattle and hogs: they eat the corn dried in the above manner as if it had lost nothing of its primitive qualities and flavor.

CHAPTER X.

THE ROOM FOR FERMENTATION.

The room destined to the fermentation must be close, lighted by two or three windows, and large enough to contain a number of hogsheads sufficient for the distillery. It may be determined by the number of days necessary for the fermentation; 30 or 40 hogsheads may suffice, each of 120 or 130 gallons.

In the middle of the room must be a stove, large enough to keep up a heat of 75° to 80°, even in winter. A thermometer placed at one end of the room, serves to regulate the heat.

As soon as the liquor is in the hogshead, the yeast, or fermenting principle, is put into it, stirred for some [Pg 26] moments, and then left to itself. A liquor as rich as the above described ferments with force, and runs with rapidity through all the periods of fermentation. It is fit to distil as soon as that tumultuous state has subsided and the liquor is calm.

The essential character of the spirituous fermentation, is to exhale the carbonic acid gaz in great quantity. This gaz is mortal to mankind, and to all the living creation. Thirty hogsheads of fermenting liquor producing a great deal of this gaz, the room should be purified of it by opening two opposite windows several times a day. This is the more essential, as the pure air, or *oxigen*, contributes to the formation of the spirit, of which it is one of the constituting principles. A short time, however, suffices to renew the air of the room.

It is useless to remark, that the hogsheads must be open at one end, and rest upon pieces of wood elevating them some inches from the ground. They must remain uncovered during the fermentation; and afterwards be covered with a flying lid, when the liquor is calm.

CHAPTER XI.

OF THE ROOM FOR DISTILLATION.

We have hitherto considered the liquor as containing only principles upon which the air has no action, and from which it can only extract some watery vapors; and, in fact, all those principles contained in the liquor are fixed. The action of the fire may concentrate, but not volatilize them. [Pg 27]

The liquor is now changed by the fermentation; it contains no longer the same principles, but has acquired those which it had not, which are volatile, and evaporate easily. They must therefore be managed carefully, in order not to lose the fruits of an already tedious labor. The spirit already created in the fermented liquor, must be collected by the distillation; but in transporting it to the still, the action of the external air must be carefully avoided, as it would cause the evaporation of some of the spirit. A pump to empty the hogsheads, and covered pipes to conduct the liquor into the still, is what has been found to answer that purpose. A good distilling apparatus is undoubtedly the most important part of a distillery. It must unite solidity, perfection in its joints, economy of fuel, rapidity of distillation, to the faculty of concentrating the spirit. Such are the ends I have proposed to myself in the following apparatus.

The usual shape of stills is defective; they are too deep, and do not present enough of surface for their contents. They require a violent fire to bring them to ebullition; the liquor at bottom burns before it is warm at the top.

My still is made upon different principles, and composed of two pieces, viz. the kettle, and its lid. The kettle, forming a long square, is like the kettle of infusion, already described, and only differs from it in being one foot deeper. The lid is in shape like an ancient bed tester; that is to say, its four corners rise into a sharp angle, and come to support a circle 16 inches diameter, bearing a vertical collar of about two inches. This collar comes to the middle of the kettle, and is elevated about 4 feet from the bottom. The lid is fas [Pg 28] tened to the kettle. The collar receives a pewter cap, to which is joined a pipe of the same metal, the diameter of which decreases progressively to a little less than 3 inches: this pipe, the direction of which is almost horizontal, is 5 feet long.

My still, thus constructed, is established upon a furnace like that of the infusion room. I observe that the side walls are only raised to the half of the height of the kettle. A vertical pipe is placed on the side opposite to the pewter one, and serves to fill up the still: it is almost at the height of the fastening of the lid, but a little above. On the same side, on a level with the bottom, is a pipe of discharge, passing across the furnace: this pipe must project enough to help to receive or to direct the fluid residue of the distillation; its diameter must be such as to operate a prompt discharge of the still.

OF THE URNS.

These are copper vessels, thus called from their resembling those funeral vases of the ancients. Mine have a bottom of about 18 inches diameter; they are two feet high, have a bulge of 6 inches near the top, and then draw in to form an overture of about 8 inches.

On one side, towards the top, there is a copper pipe 2 inches diameter, projecting externally 2 or 3 inches, and bent in an elbow: it enters the internal part of the urn, and descends towards the bottom, without touching it; there it is only a slight curve, and remains open.

The external part of that pipe is fitted to receive the pewter pipe of the still; they are made so as to enter into one another, and must fit exactly. The round opening at the top of the urn receives a cap with a pewter pipe, made like that of the still. It is likewise five [Pg 29] feet long, and its size in proportion to the opening: this goes and joins itself to the second urn, as the still does to the first. The pipe of this second goes to a third, and the pipe of this last to the worm. The three urns bear each a small pipe of discharge towards the bottom.

This apparatus must be made with the greatest care. Neither the joints, the different pipes of communication, nor the nailings, must leave the smallest passage to the vapors. The workman must pay the greatest attention to his work, and the distiller must lute exactly all the parts of the apparatus that are susceptible of it: he must be the more careful as to luting it, as this operation is only performed once a week, when the apparatus is cleaned. At the moment of the distillation, the master or his foreman must carefully observe whether there is any waste of vapors, and remedy it instantly. The still and urns ought to be well tinned.

CHAPTER XII.

EFFECTS OF THIS APPARATUS.

Although the still might contain 400 gallons, there must be only 200 gallons put into it: the rest remaining empty, the vapors develops themselves, and rise. In that state, the vinous liquor is about one foot deep, on a surface of 20 feet square: hence two advantages—the first, that being so shallow, it requires but little fuel to boil; the second, that the extent of surface gives rise to a rapid evaporation, which accelerates the work. This acceleration is such, that six distillations might be obtained in one day. The spirit contained in the vinous liquor rises in vapors to the lid of the still, [Pg 30] there find the cap and its pipe, through which they escape into the first urn, by the side pipe above described, which conducts them to the bottom, where they are condensed immediately.

But the vapors, continuing to come into the urn, heat it progressively: the spirituous liquor that it contains rises anew into vapors, escapes through the cap and pipe, and arrives into the second urn, where it is condensed as in the first. Here again, the same cause produces the same effect: the affluence of the heat drawn with the vapors, carries them successively into the third urn, and from thence into the worm, which condenses them by the effects of the cold water in which it is immersed.

The urns, receiving no other heat than that which the vapors coming out of the still can transmit to them, raise the spirit; the water, at least the greatest part of it, remains at the bottom: hence, what runs from the worm is alcohol; that is, spirit at 35°. It is easily understood how the vapors coming out of the still are rectified in the urns, and that three successive rectifications bring the spirit to a high degree of concentration: it gets lower only when the vinous liquor draws towards the end of the distillation. As soon as it yields no more spirit, the fire is stopped, and the still is emptied in order to fill it up again, to begin a new distillation.

Each time that the vinous liquor is renewed in the still, the water contained in the urns must be emptied, through the pipes of discharge at the bottom.

Metals are conductors of the *caloric*. The heat accumulated in the still, rises to the cap, from whence it runs into the urns: with this difference—that the pewter, of which the cap and pipes are made, transmits less [Pg 31] caloric than copper, because it is less dense: and that bodies are only heated in reason of their density.

However, a great deal of heat is still communicated to the worm, and heats the water in which it is immersed. I diminish this inconvenience by putting a wooden pipe between the worm and the pipe of the third urn. Wood being a bad conductor of caloric, produces a *solution of continuity*, or interruption between the metals. The wood of this pipe must be soft and porous, and not apt to work by the action of the fire: however, to avoid its splitting, I wrap it up in two or three doubles of good paper, well pasted, and dried slowly. This pipe is one foot long, and hollowed in its length, so as to receive the pewter pipe of the third urn at one end, and to enter the worm at the other; thereby the worm is not as hot, since it only receives the heat of the vapors which it condenses.

Notwithstanding all these precautions, it heats the water in which it is immersed after a length of time; and whatever care may be taken to renew it, all the vapors are not condensed, and this occasions a loss of spirit. I obviate this accident, by adding a second worm to the first: they communicate by means of a wooden pipe like the above. The effect of this second worm, rather smaller than the first, is such, that the water in which it is plunged remains cold, while that of the first must be renewed very often. By these means, no portion of vapors escape condensation. The liquor running from the worm is received into a small barrel, care being taken that it may not lose by the contact of the air producing evaporation. [Pg 32]

CHAPTER XIII.

OF FERMENTS.

They are of two kinds; the very putrescent bodies, and those supplied by the *oxigen*. Animal substances are of the first kind: *acids*, neutral salts, rancid oils, and metallic *oxids*, are of the second.

Were I obliged to make use of a ferment of the first class, I would choose the glutinous part of wheat flour. This vegeto-animal substance is formed in the following manner:—A certain quantity of flour is made into a solid dough, with a little water. It is then taken into the hands, and water slowly poured over it, while it is kneaded again. The water runs white, because it carries off the starchy part of the flour; it runs clear after it is washed sufficiently. There remains in the hands of the operator a dough, compact, solid, elastic, and reduced to nearly the half of the flour employed. This dough, a little diluted with water, and kept in the temperature indicated for the room of fermentation, passes to the putrid state, and contracts the smell of spoiled meat. Four pounds of this dough per hogshead, seem to me to be sufficient to establish a good fermentation. A small quantity of good vinegar would answer the same purpose, and is a ferment of the second class.

But are those means indispensable with my process? I do not think so.

1st. The richness of my vinous liquor, and the degree of heat to which I keep it, tend strongly to make it ferment. In fact, the infusion of the grain, by taking from it its saccharine part, takes likewise part of its mucilaginous substance, which is the principle of the [Pg 33] spirituous fermentation, which it establishes whenever it meets with the other substance.

2dly. The hogsheads themselves are soon impregnated with a fermenting principle, and communicate it to the liquor that is put into them.

3dly. The rum distiller employs advantageously the residue of his preceding distillation, to give a fermentation to his new molasses: this residue has within itself enough of acidity for that purpose.

Might not the residue of the distillation of my vinous liquor have the same acidity? It contains only the mucilaginous substance already acidulated. Some gallons of that residue to every hogshead, would, I think, be a very good ferment.

Lastly. Here is another means which will certainly succeed: it is to leave at the bottom of each hogshead three or four inches of the vinous liquor, when transported into the still for distilling. This rising, which will rapidly turn sour, will form a ferment sufficient to establish a good fermentation.

The intelligent manager of a distillery must conduct the means I indicate, towards the end which he proposes to himself, and must carefully avoid to employ as ferments, those disgusting substances which cannot fail to bring a discredit on the liquor in which they are known to be employed.

CHAPTER XIV.

OF THE AREOMETER, OR PROOF BOTTLE.

This instrument is indispensable to the distiller: it ascertains the value of his spirits, since it shows the [Pg 34] result of their different degrees of concentration. I will give the theory of this useful instrument, as it may be acceptable to those who do not know it.

Bodies sink in fluids, in a *compound ratio* to the volume and the density of those fluids, which they displace. It is from that law of nature, that a ship sinks 20 feet in fresh water, while it sinks only about 18 feet in sea water, which has more density on account of the salt dissolved therein.

The reverse of this effect takes place in fluids lighter than water, as bodies floating in them sink the more, as the liquor has less density. Upon those principles are made two kinds of areometers—one for fluids denser than water; the other for those that are lighter: the first are called *salt proof*; the second *spirit proof*. Distilled water is the basis of those two scales: it is at the top for the *salt proof*, and at the bottom for the *spirit proof*; because the first is ascending, and the other descending; but by a useless singularity, the distilled water has been graduated at 10° for the spirit proof bottle, and at 0 for the *salt proof*. We shall only dwell upon the first, because it is the only one interesting to the distiller.

Water being graduated at 10° in the areometer, it results from thence that the spirit going to 20°, is in reality only 10° lighter than water; and the alcohol graduated at 35°, is only 25° above distilled water.

The areometer can only be just, when the atmosphere is temperate; that is, at 55° Fahrenheit, or 10° Reaumur. The variations in cold or heat influence liquors; they acquire density in the cold, and lose it in the heat: hence follows that the areometer does not [Pg 35] sink enough in the winter, and sinks too much in the summer.

Naturalists have observed that variation, and regulated it. They have ascertained that 1° of heat above temperate, according to the scale of Reaumur, sinks the areometer 1/8 of a degree more; and that 1° less of heat, had the contrary effect: thus the heat being at 18° of Reaumur, the spirit marking 21° by the areometer, is really only at 20°. The cold being at 8° below temperate, the spirit marking only 19° by the areometer, is in reality at 20°. 2¼ of Fahrenheit corresponding to 1° of Reaumur, occasion in like manner a variation of 1/8 of a degree: thus, the heat being at 78½°, the spirit thus marking 21°, is only at 20; and the cold being at 87°, the spirit marking only 19° by the areometer, is in reality at 20°.

It is easily conceived, that extreme cold or extreme heat occasion important variations. For that reason, there are in Europe inspectors, whose duty it is to weigh spirits, particularly *brandy*: for that purpose they make use of the areometer and the thermometer. An areometer, to be good, must be proved with distilled water, at the temperature of 55°. Areometers, being made of glass, are brittle, and must be used with great care. This inconvenience might be remedied, by making them of silver; I have seen several of this metal. A good silversmith could easily make them; I invite those artists to attend to that branch of business; it might become valuable, as the distillers will be more enlightened. [Pg 36]

CHAPTER XV.

ADVANTAGES OF MY METHOD.

The first of all, is derived from the composition of a vinous liquor, richer, and more proper to raise a vigorous fermentation, than that which is obtained by the usual method. Now, as it is proved that the quantity of spirit is in proportion to the richness of the fermenting liquor, mine therefore yields a great deal more spirit than any other.

2dly. We have seen that a heat of 75° or 80° must be kept up in the fermenting room: this being summer heat, proves that such a rich vinous liquor runs no risk of passing to the acid state with as much rapidity as that of the common distillers; and, consequently, that he who will follow my method can work all the year round without fear of losing the fruits of his labor, as it often happens—an advantage precious for him who makes it his sole business. The only change he has to make, is to suppress the heat of the stove, when the temperature of the atmosphere is sufficient to keep up a good fermentation in the liquor.

As to my distilling apparatus, this is not a new idea. I present it to the public under the sanction of experience. I had it executed in Philadelphia eight years ago, after having obtained a patent. It was made for a rum distillery, where they still continue to use it. It presents the greatest advantages.

The first is, that with a single fire, and a single workman, I distil and rectify the spirit three times, and bring it to the degree of alcohol; that is, to the greatest purity, and almost to the highest degree of concentration. [Pg 37]

2dly. It lowers the cost of transportation, by two-thirds; because one gallon at 35° represents three gallons at the usual degree. The merchant, being arrived at the place of his destination, has only to add 2 gallons of water to 1 gallon of this alcohol, in order to have 3 gallons of whiskey; which is of a considerable advantage, either for land or sea carriage.

3dly. As the price of spirits is, in trade, in proportion to their degree of concentration, those made with my apparatus being at a very high degree, need no more rectifying, either for the retailer, the apothecary, or the painter; and the considerable expenses of that operation turn entirely to the profit of the distiller, as they are totally suppressed. Distillers may hereafter sell spirits of all degrees of concentration.

Such are the advantages of my processes. I offer them the more willingly to the public, as they are founded upon the most approved principles of natural philosophy: by reflecting upon them, distillers will be easily convinced of it.

However perfect the description of a new thing may be, our ideas of it are always defective, until we have seen it put into practical use. Few men have the means of establishing a distillery on a new plan, and even the most enlightened may make notable errors. Few, besides, are bold enough to undertake, at their own risks, the trial of a new fabrication: they are afraid of losing, and of being blamed for having too lightly yielded to the persuasion of new projectors. Hence it follows that a useful discovery falls into oblivion, instead of doing any good. [Pg 38]

But no discovery of general utility ought to experience that fate in a republic. Government itself ought to promote the first undertaking, or a certain number of citizens ought to join in order to give it a start. It is the more easy in this case, as my apparatus requires very little expense.

If a distillery according to my directions, was established in some of the principal towns of the state, my method would then make rapid progress, and thus prove the truth of the principle which I have advanced; and the distillers, after having meditated upon my method in this book, would come and satisfy themselves of its goodness, by seeing it put into practice, and yielding the most perfect results, with all the advantages for trade that may be expected: hence would naturally ensue the rapid increase of distillation, and consequently that of agriculture and commerce. [Pg 39]

THE ART OF MAKING GIN, AFTER THE PROCESS OF THE HOLLAND DISTILLERS.

Having indicated the most proper means of obtaining spirits, I will now offer to the public the manner of making *Gin*, according to the methods used by the distillers in Holland. It may be more properly joined to the art of making whiskey, as it adds only to the price of the liquor, that of the juniper berries, the product of which will amply repay its cost. Many distillers in the United States have tried to imitate the excellent liquor coming from Holland, under the name *gin*. They have imagined different methods of proceeding, and have more or less attained their end. I have myself tried it, and my method is consigned in a patent.

But those imitations are far from the degree of perfection of the Holland gin: they want that unity of taste, which is the result of a single creation; they are visibly compounds, more or less well combined, and not the result of a spontaneous production.

To this capital defect, which makes those imitations so widely different from their original, is joined their high price, which prevents its general consumption. In fact, it is made at a considerable expense: the whiskey must be purchased, rectified and distilled over again [Pg 40] with the berries. These expenses are increased by the waste of spirit occasioned by those reiterated distillations. This brings the price of this false gin to three times that of the whiskey: consequently the poorer sort of people, whose number is always considerable, are deprived of the benefits of a wholesome liquor, and restrained to whiskey, which is commonly not so.

The methods used in Holland, have reduced gin to the lowest price; that of the juniper berries being there very trifling, and increasing but little the price of whiskey: still that small addition is almost reduced to nothing, as will be seen hereafter.

The United States are, in some parts, almost covered with the tree called here *cedar*; which tree is no other than the juniper, and grows almost every where, and bears yearly a berry, which is in reality the juniper berry. Some Hollanders knew it at Boston, collected considerable quantities of it in Massachusetts, and shipping it to some of the eastern harbors, sold it as coming from Holland. I have seen some at Philadelphia ten years ago, at the house of a Hollander, who received it from Massachusetts in hogsheads of about ten hundred weight, and sold as the produce of his own country, what was really that of the United States.

I collected myself a great quantity of those berries, at Norfolk, Va. by means of negroes, to whom I paid one dollar per bushel of 40 lbs. being 2½ cts. per pound. Two years ago, it sold for 6 cents in Philadelphia, and bore the same price at Pittsburgh.

There is a great deal of cedar in Kentucky, and consequently of berries. I have seen them at Blue Licks, and they abound near the Kentucky river. [Pg 41]

Although an incredible number of those trees is cut down daily, there is still a greater number standing, in the United States; and millions of bushels of berries are lost every year, while only skilful hands are wanted, to make them useful to mankind. The juniper berry has many medical properties: it is a delightful aromatic, and contains an oil essential, and a sweet extract, which by the fermentation yields a vinous liquor, made into a sort of wine in some countries; that is called wine for the poor: it strengthens the stomach, when debilitated by bad food or too hard labor.

The Hollanders, who have long had the art of trading upon every thing, have constantly turned even their poverty to account. They have immense fabrications of gin, and scarcely any juniper trees. They only collect the berry in those countries where it is neglected as useless, as in France and Tyrol, which produce a great deal of it. The United States need have no recourse to Europe, in order to get the juniper berries: they have in abundance at home, what the Hollanders can only procure with trouble and money. They can therefore rival them with great advantage; but they must follow the same methods employed in the Holland distilleries.

The juniper berry contains the sweet mucous extract, in a great proportion: it has therefore the principle necessary to the spirituous fermentation; and, indeed, it ferments spontaneously. When fresh, and heaped up, it acquires a degree of heat, but not enough to burn, as I have ascertained: it is therefore safely transported in hogsheads. From that facility of fermenting, it must be considered as a good ferment, and as increasing the quantity of spirit, when joined to a fermentable liquor. [Pg 42]

A distiller may at pleasure convert his whiskey into gin. He needs only to perfume the wort which he puts in fermentation, by adding a certain quantity of the berries, slightly broken: the fermentation is then common to both; their sweet mucosity enriches that of the wort, and increases the spirit, while at the same time the soapy extract, which is the proximate principle of vegetation, yields the essential oil, which perfumes the liquor. [c]

The fermentation being common to both substances, unites them intimately; and when, by the distillation, the spirit is separated from the water, there remains an homogenous liquor, resulting from a single creation, and having that unity of taste, and all the properties of Holland gin, because obtained by the same means.

One single and same distillation can therefore yield to the distiller either gin or whiskey, as it requires no more labor, and its conversion into gin costs only the price of the berries, which repays him amply, either by the spirit it yields, or by its essential oil, which, floating on the surface, may be easily collected. This oil bears a great price, and the Hollanders sell much of it.

We have seen, in the 10th chapter of this work, that my hogsheads for the fermentation, contain about 120 gallons of wort, being the production of the saccharine extract of 12 bushels of grain. The intelligent distiller will himself determine the quantity of berries necessary for each hogshead to have a good aromatic per [Pg 43] fume. He may begin with 10 lbs. per hogshead; and will, upon trial, judge whether or not this quantity is sufficient, or must be increased. At any rate, economy should not be consulted in the use of the berries, since their price does not increase that of the whiskey. This low price must naturally become the principle of an immense fabrication of gin; and henceforth it will be an important article of exportation for the United States, as well as a considerable and wholesome object of home consumption.

Footnotes:

[a] Some rum distillers make a stronger vinous liquor, but it is still very far from Lavoisier's proportions. Others add successively new molasses to their vinous liquor, and thus prolong their fermentation, without making their liquor stronger, and consequently without obtaining more spirit. This is absolutely contrary to the true principles of distillation.

[b] See his beautified operation on the decomposition of water.

[c] I must here observe, that the juniper berry, as well as several other fruits, contains two kinds of essential oil: one is the proximate principle of vegetation, and the other is the superabundant oil: the first is combined with the soapy extract, and dissolves in water; while the second does not unite with it, and floats on the surface.

END